How to Make a Racing Car

Debbie Croft

Photographs by
Lindsay Edwards

Contents

Goal

To make a racing car

Materials

You will need:

- glue

- paints

- light card (15 cm × 15 cm)

- paintbrushes

- scissors

- a pencil

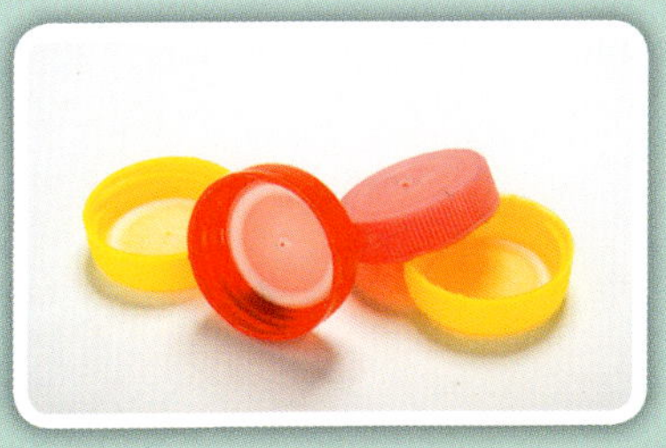

- four plastic lids, each with a hole in the middle

- two plastic straws, each one 5 cm long

- two toothpicks, each one 6 cm long

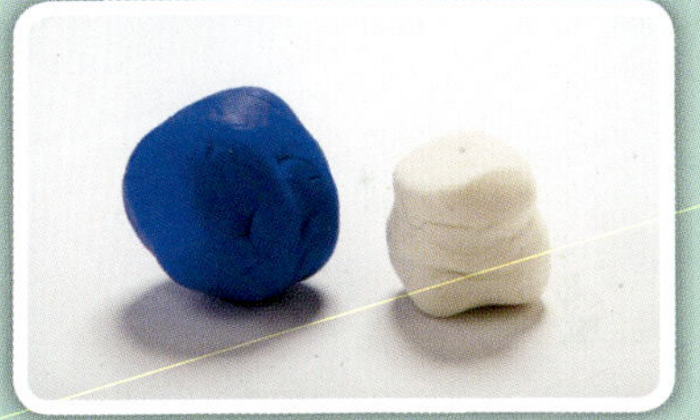

- plasticine

- shiny paper

- a plastic racing-car driver.

Steps

Make the Body of the Racing Car

1. Place the card on the table.

Fold the card down the middle.

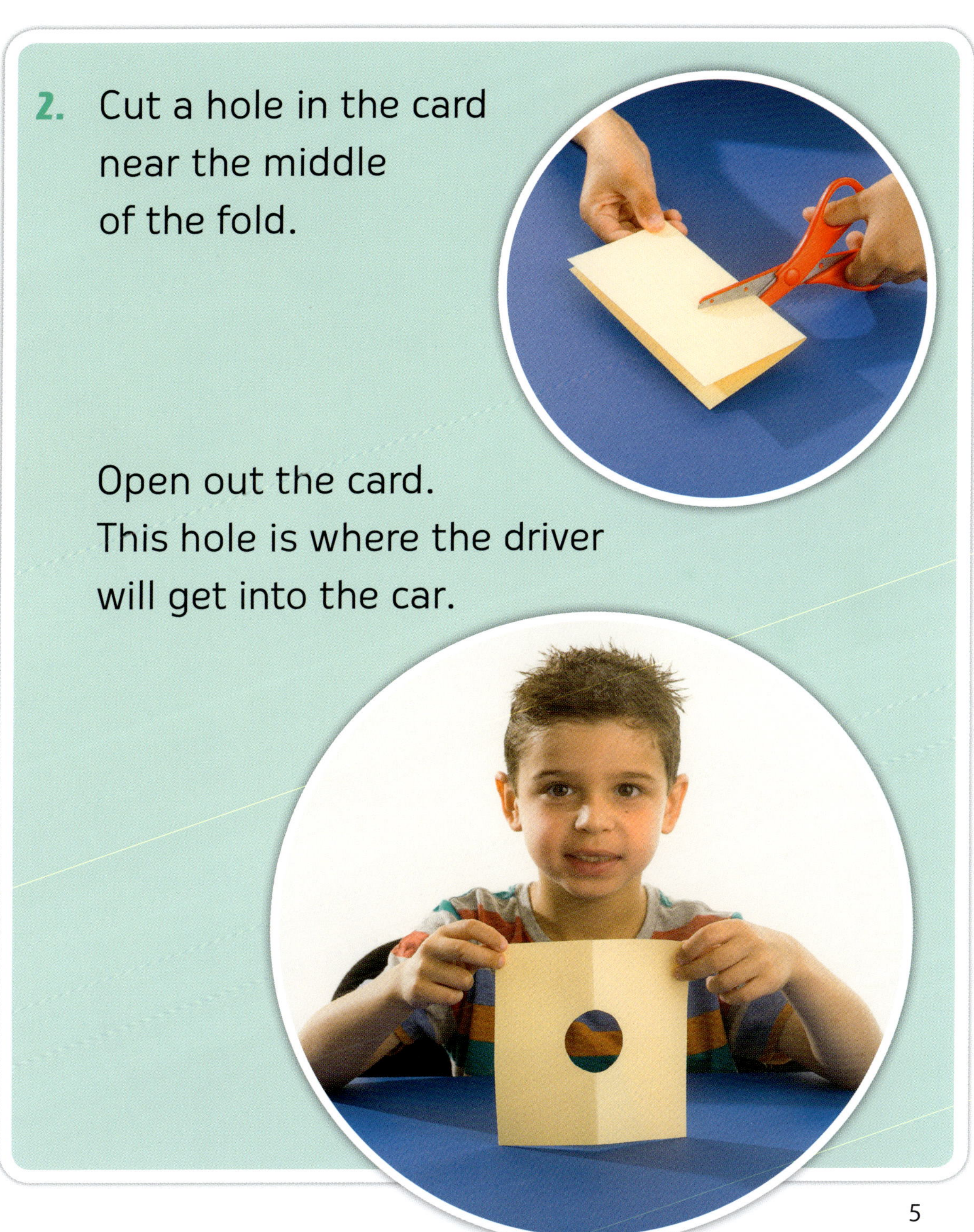

2. Cut a hole in the card near the middle of the fold.

Open out the card.
This hole is where the driver will get into the car.

3. Put some glue along one side of the card.

Make the card into a long **cylinder**. Glue the sides together.

This cylinder will be the body of the racing car.

Paint the Racing Car and the Wheels

4. Paint the body
of the racing car
with your favourite colours.

Paint a number on the front
of the car.
Paint this number
on each side
of the car, too.

5. Paint the four plastic lids black.
These will be the wheels on the car.

Put the Wheels on the Racing Car

6. Make a hole
   on each side of the cylinder
   with the pencil.
   These holes will be for the front wheels.

7. Push a straw into one hole
   until it comes out the hole
   on the other side.

8. Put a toothpick into the hole in one of the lids. This makes the front **axle**.

Put some plasticine on the outside of the lid. This will stop it from falling off.

9. Put the toothpick into the straw at the front of the car.
Then, put another lid on the other end of the toothpick.
Put some plasticine on the outside of this lid, too.

10. Do steps 6 to 9 again. This will make the wheels for the back of the car.

11. Put some shiny paper inside the lid on each wheel.

Make a Seat for the Driver

12. Make a **seat** for the driver out of plasticine.

Push the seat into the car so it does not move.

Put the driver in his seat in the car.

Now the car is ready to race!

Glossary

axle a rod between two wheels

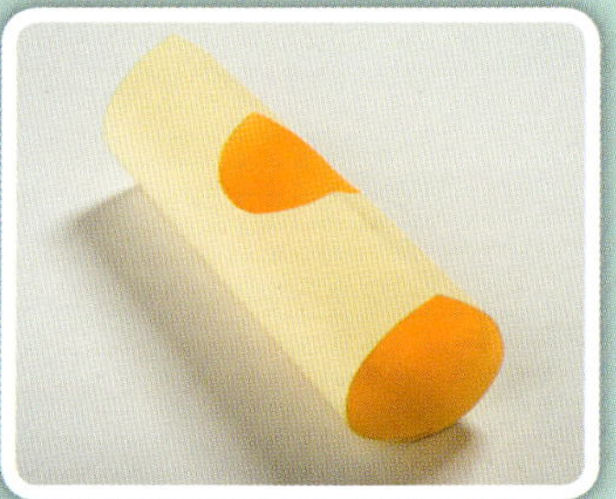

cylinder a long, round shape

fold to bend over

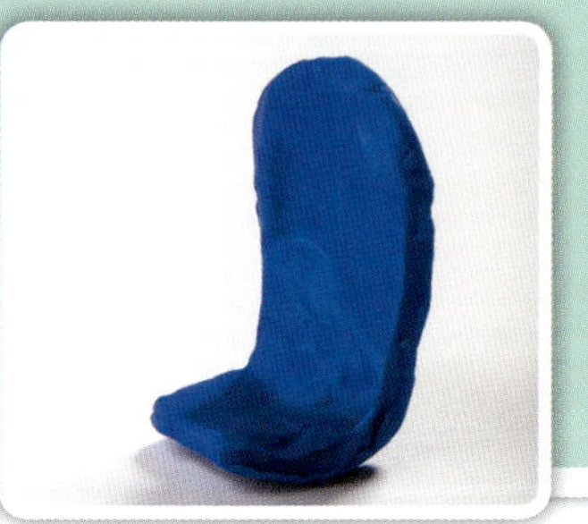

seat a place to sit